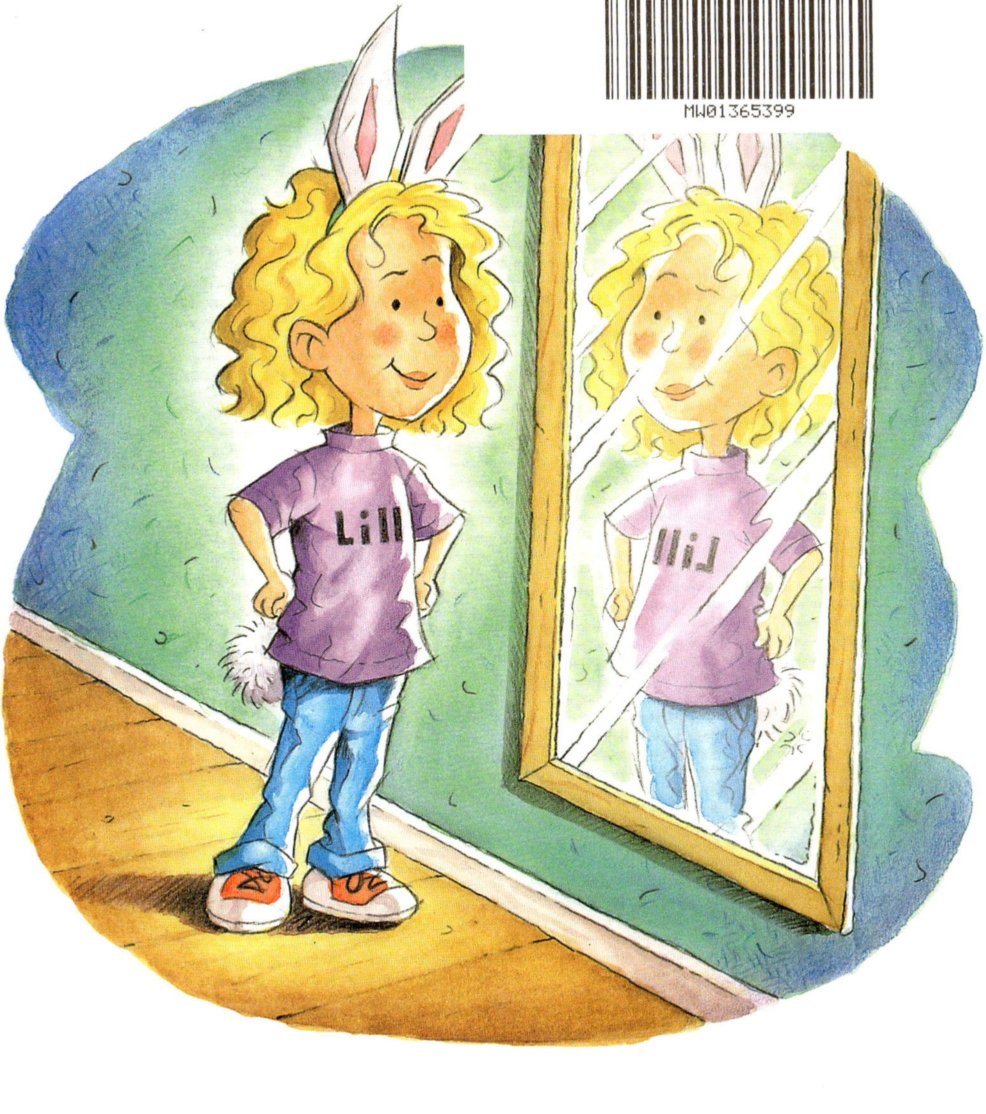

I am not Lill.

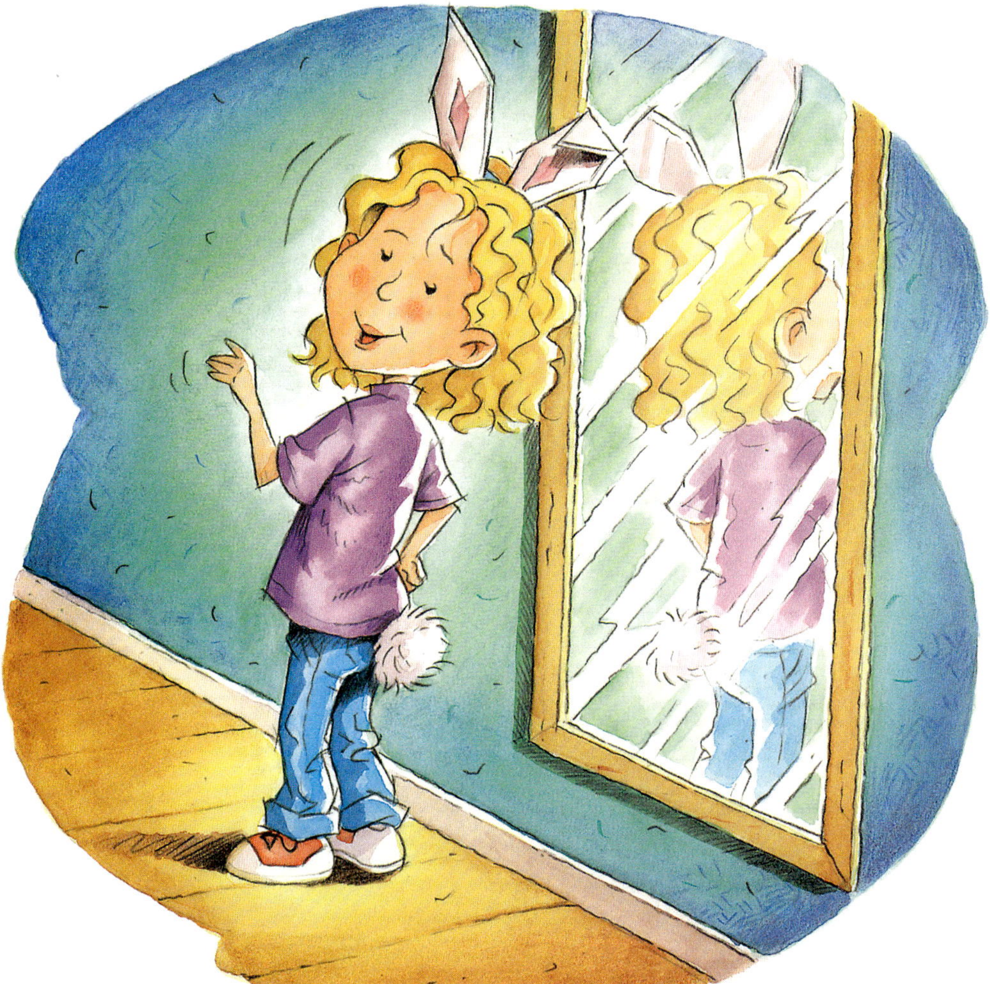

I am a 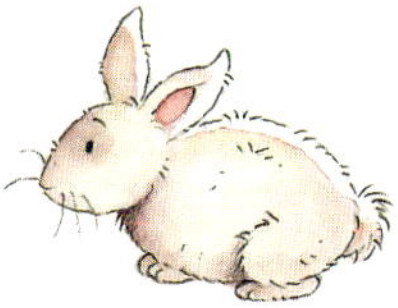 !

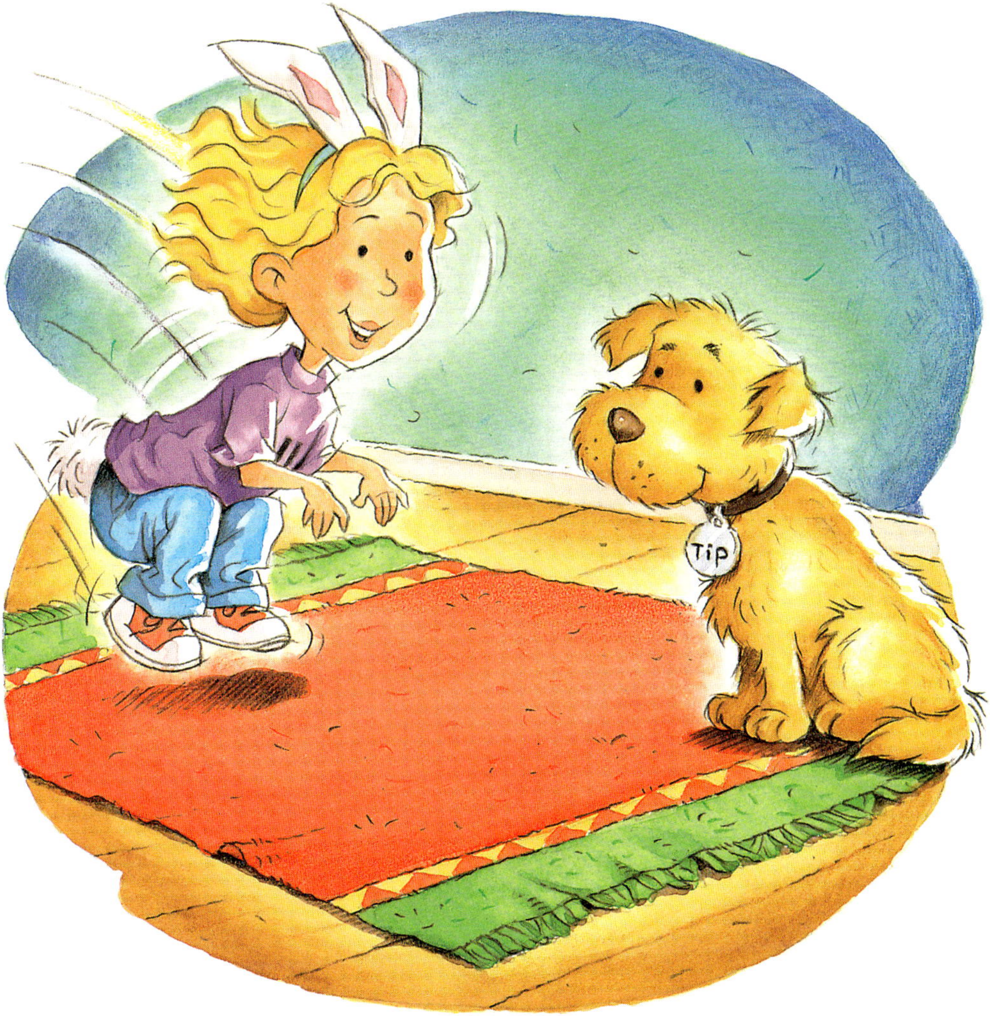

I hop on a mat to Tip.

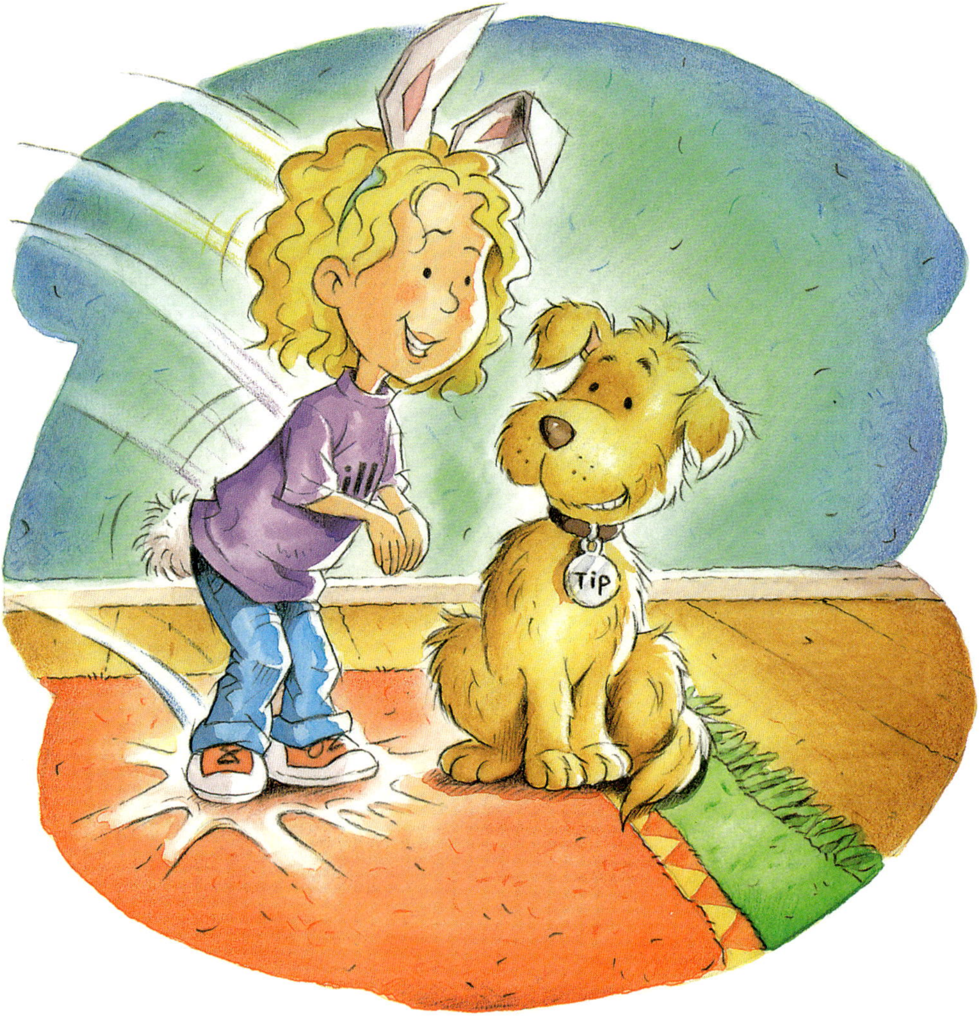

I hop to him.

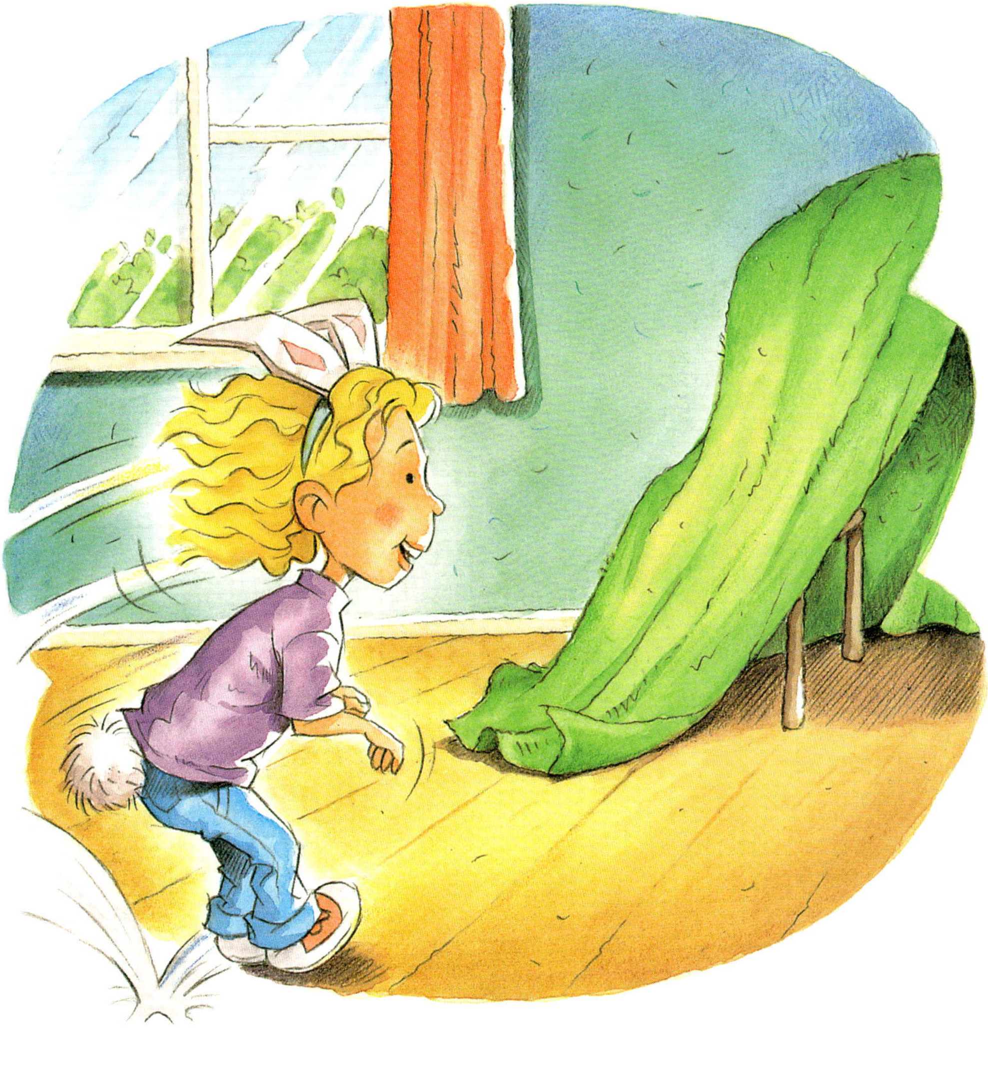

I hop to a hill.

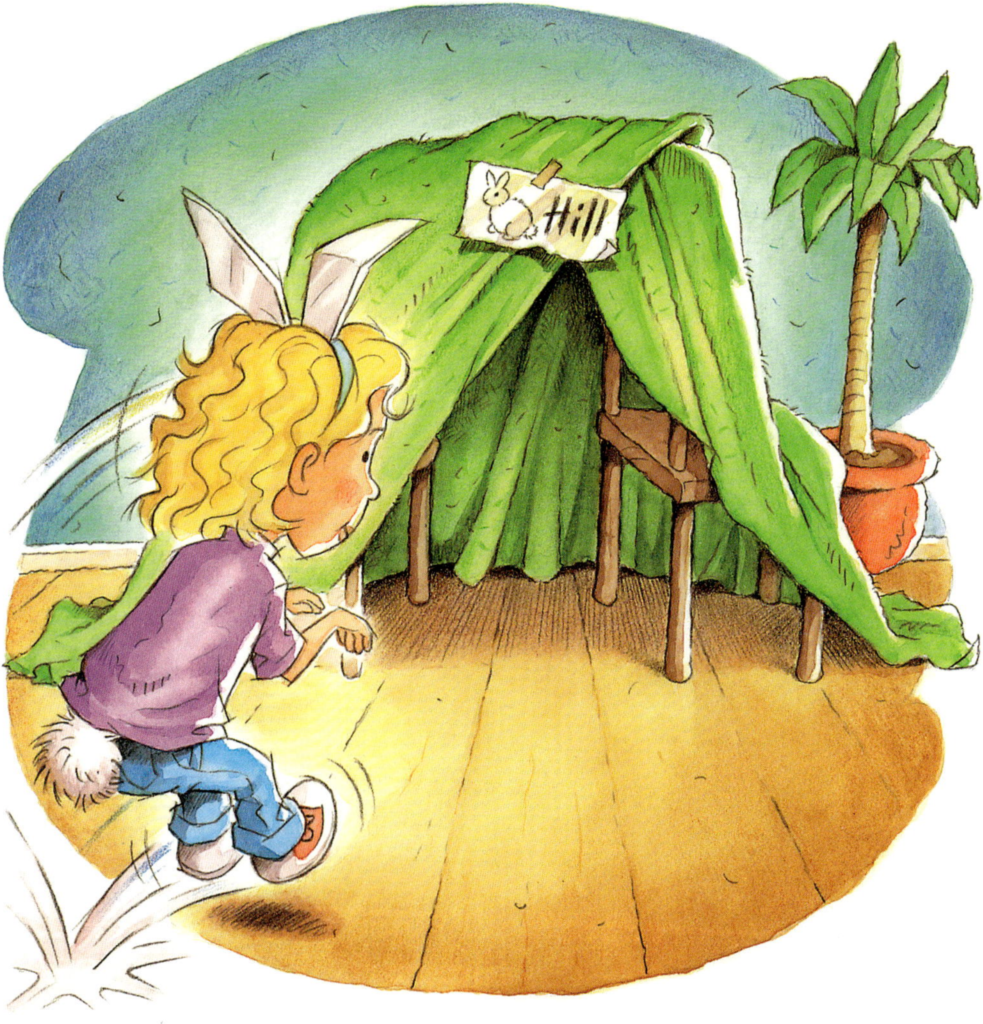

I hop, hop to it.

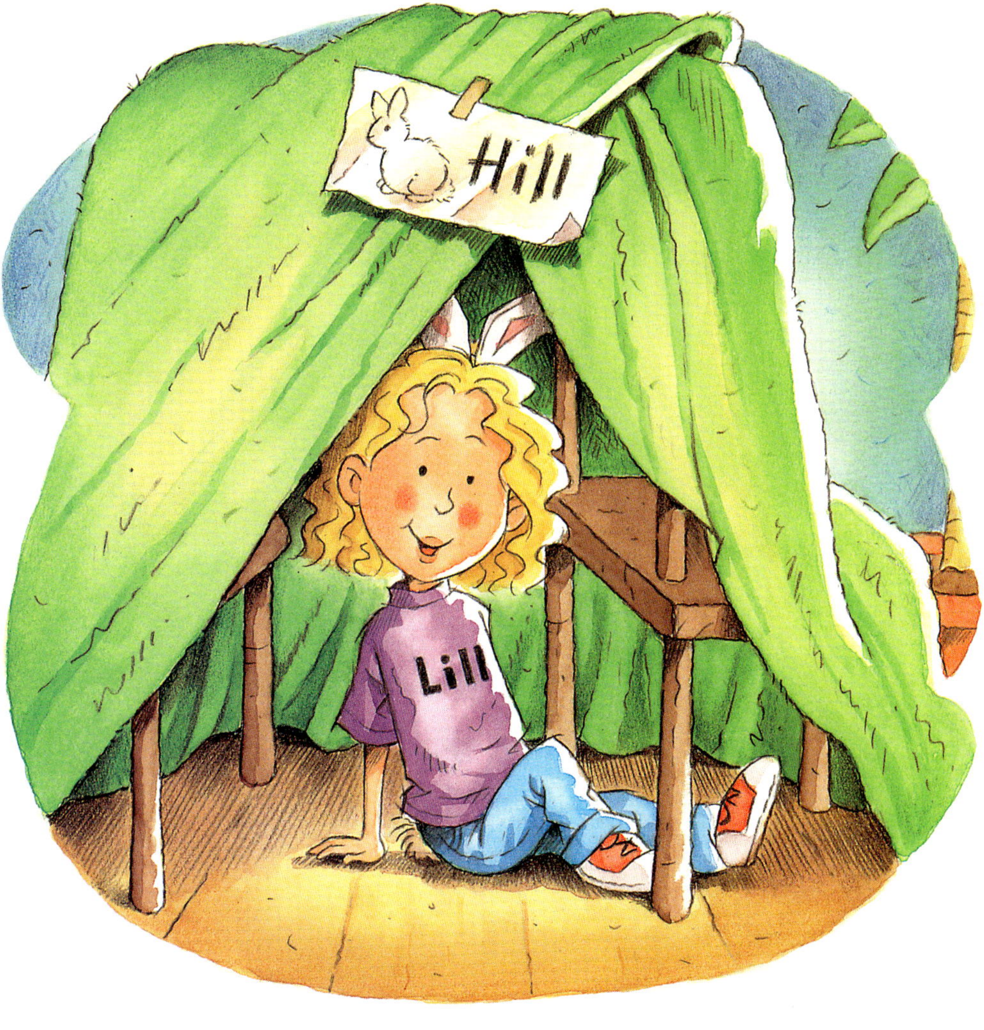

I am in the hill.

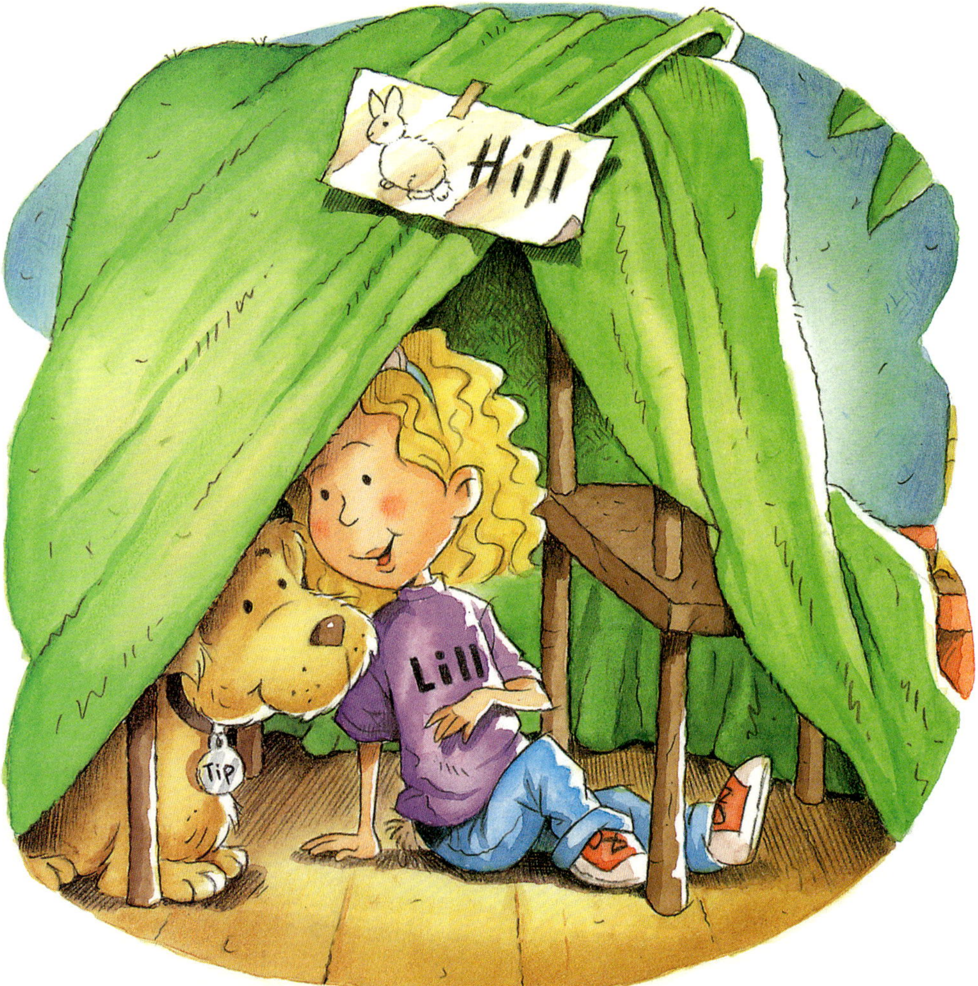
Hill
Lill
Tip

Hop in, Tip!